AF879701

Strategic Studies Institute
and
U.S. Army War College Press

THUCYDIDES WAS RIGHT: DEFINING THE FUTURE THREAT

Colin S. Gray

April 2015

The views expressed in this report are those of the author and do not necessarily reflect the official policy or position of the Department of the Army, the Department of Defense, or the U.S. Government. Authors of Strategic Studies Institute (SSI) and U.S. Army War College (USAWC) Press publications enjoy full academic freedom, provided they do not disclose classified information, jeopardize operations security, or misrepresent official U.S. policy. Such academic freedom empowers them to offer new and sometimes controversial perspectives in the interest of furthering debate on key issues. This report is cleared for public release; distribution is unlimited.

*****

This publication is subject to Title 17, United States Code, Sections 101 and 105. It is in the public domain and may not be copyrighted.

*****

Comments pertaining to this report are invited and should be forwarded to: Director, Strategic Studies Institute and U.S. Army War College Press, U.S. Army War College, 47 Ashburn Drive, Carlisle, PA 17013-5010.

*****

This manuscript was funded by the U.S. Army War College External Research Associates Program. Information on this program is available on our website, *www.StrategicStudiesInstitute.army.mil,* at the Opportunities tab.

*****

All Strategic Studies Institute (SSI) and U.S. Army War College (USAWC) Press publications may be downloaded free of charge from the SSI website. Hard copies of this report may also be obtained free of charge while supplies last by placing an order on the SSI website. SSI publications may be quoted or reprinted in part or in full with permission and appropriate credit given to the U.S. Army Strategic Studies Institute and U.S. Army War College Press, U.S. Army War College, Carlisle, PA. Contact SSI by visiting our website at the following address: *www.StrategicStudiesInstitute.army.mil.*

*****

The Strategic Studies Institute and U.S. Army War College Press publishes a monthly email newsletter to update the national security community on the research of our analysts, recent and forthcoming publications, and upcoming conferences sponsored by the Institute. Each newsletter also provides a strategic commentary by one of our research analysts. If you are interested in receiving this newsletter, please subscribe on the SSI website at *www.StrategicStudiesInstitute.army.mil/newsletter.*

ISBN 1-58487-670-0

FOREWORD

To define future threat is, in a sense, an impossible task, yet it is one that must be done. In this monograph, Dr. Colin S. Gray explains that the only sources of empirical evidence accessible to us are the past and the present. We cannot obtain understanding about the future from the future.

Dr. Gray draws noticeably upon the understanding of strategic history obtainable from Thucydides' great *History of the Peloponnesian War*. The monograph advises prudence as the operating light for American definition of future threat, and the author believes that there are historical parallels between the time of Thucydides and our own that can help us avoid much peril. The future must always be unpredictable to us in any detail, but the many and potent continuities in history's great stream of time can serve to alert us to what may well happen in kind.

The Strategic Studies Institute is pleased to offer this monograph to assist policymakers and strategic thinkers in understanding this vital topic.

DOUGLAS C. LOVELACE, JR.
Director
Strategic Studies Institute and
U.S. Army War College Press

ABOUT THE AUTHOR

COLIN S. GRAY is Professor of International Politics and Strategic Studies at the University of Reading, England. He worked at the International Institute for Strategic Studies in London, UK, and at the Hudson Institute in Croton-on-Hudson, NY, before founding the National Institute for Public Policy, a defense-oriented think tank in the Washington, DC, area. Dr. Gray served for 5 years in the Ronald Reagan administration on the President's General Advisory Committee on Arms Control and Disarmament. A dual citizen of the United States and UK, he has served as an adviser to both the U.S. and British governments. His government work has included studies of nuclear strategy, arms control, maritime strategy, space strategy, and special forces. Dr. Gray has written 27 books, including: *The Sheriff: America's Defense of the New World Order* (University Press of Kentucky, 2004); *Another Bloody Century: Future Warfare* (Weidenfeld and Nicolson, 2005); *Strategy and History: Essays on Theory and Practice* (Routledge, 2006); *Fighting Talk: Forty Maxims on War, Peace and Strategy* (Potomac Books, 2009); *National Security Dilemmas: Challenges and Opportunities* (Potomac Books, 2009); *The Strategy Bridge: Theory for Practice* (Oxford University Press, 2010); *War, Peace and International Relations: An Introduction to Strategic History*, 2nd Ed. (Routledge, 2011); *Airpower for Strategic Effect* (Air University Press, 2012); *Perspectives on Strategy* (Oxford University Press [OUP] 2013), which is the follow-on to *Strategy Bridge*. The final volume in the *Strategy Bridge* trilogy, entitled *Strategy and Defence Planning: Meeting the Challenge of Uncertainty*, was published by OUP in 2014. Dr. Gray is a graduate of the Universities of Manchester and Oxford.

SUMMARY

This monograph examines the challenge in future threat definition. In order to do so, it is necessary to understand where identification of threat originates, and how and why such identification is made in the context of international political relations. This analysis makes fairly heavy use of the ideas in Thucydides' great *History of the Peloponnesian War*. Effort is expended here to explain why a work written in Greece, in the late-5th century B.C., has high value for us today as a vital aid to understanding of our own current, and indeed future, security context.

The monograph offers conclusions/recommendations in four broad clusters. First, prudence is recommended as the guiding light in the face of an irreducible ignorance about the future. Second, the monograph explains that there is considerable real (political and cultural) discretion about the particular identification and definition of threat. Third, the analysis flatly rejects the idea of historical analogy as a vital source of evidence on future threat; instead, the concept of the historical parallel, the difference between the two ideas, is very large. Fourth, although the contemporary United States is indeed unique and exceptional as an actor on the world state, it is nonetheless simply a very large and powerful state that is obliged to behave according to the same rule book, and plan with a familiar playbook, as have other great powers of the past and present.

THUCYDIDES WAS RIGHT: DEFINING THE FUTURE THREAT

> The fact that, historically speaking, correct threat perception is exceptional—resting perhaps on luck or intuitive judgement, that is, on unreliable resources—stands in sharp contrast to the common assumption that threat perception is easy.
>
> Klaus Knorr, 1976[1]

> Yet the only empirical data we have about how people conduct war and behave under its stresses is our experience with it in the past, however much we have to make adjustments for subsequent changes in conditions.
>
> Bernard Brodie, 1976[2]

WHAT IS THE PROBLEM?

The subject that lies unavoidably at the core of this analysis is the challenge of understanding both what Americans will, or certainly should, understand about threat in the future and what kinds of policy and strategy actions might be implied as appropriate. Unfortunately, elements key to this inquiry are far from being all but self-explanatory, even when or if the familiarity of popular concepts appears to attribute a quite undeserved merit to them. By way of firing two "shots across the bow" of this subject and accepting the risk of my appearing unduly negative, it is necessary to decline to be unduly impressed by the two highly potent ideas that must drive this study—threat and the future. The sheer familiarity of these concepts can paralyze our critical capacity. I should hasten to explain that the high purpose here does not lie in the exploration of interesting ideas in a quest after understanding for its own sake. Rather, this inquiry is motivated and organized with the intention of discovering a way, or ways, in which the United States, and particularly its Army, could act in order to counter the problems that attend and infest the concepts of the future and of threat. The fundamental subject is the challenge to the United States that inalienably is issued by the need to be able to cope well enough with what may come to be regarded in the future as threats.

Notwithstanding the uncontentious linked dual concepts of threat and the future, it is necessary to recognize that both concepts are characterized by immense contemporary lack of knowledge on our part. The beginning of wisdom has to be frank recognition of our total ignorance about detail of and from the future. This is an obvious and unarguable truth that carries profound meaning for politics and strategy, but all too rarely is properly understood. In fact, so truly resistant are people to the full meaning of the "future" that often they succumb to the temptation in effect to deny temporal actuality. The somewhat awkward truths are that we can obtain no information about the future from the future, and indeed that the future never arrives.[3] Officials and scholars do not challenge this law of physics, but they are prone to forget it, or at least to behave in their endeavors in planning as if it were only a matter of minor inconvenience. In the analysis and discussion here, I attempt to address the awesomely challenging problem of how the United States and its Army can be ready enough to cope adequately with the difficulty of not knowing for certain what it may be required to do, or against whom it may have to be prepared to do it. It is necessary to begin by acknowledging that any, and indeed every, explanation of future threat has to be able to refer to a body of evidence in its regard. What unlocks somewhat the mystery of future threat can only live in the evidence we have more or less accessible to us, indirectly, in the record of strategic history, past and present.

We are used to employing the apparently opposed, though actually complementary, ideas of continuity and change. To enrich greatly this dynamic duo, we need also to contextualize treatment of future threat by adopting the master notion of the great stream of time.[4] This imperially inclusive organizing concept requires us to accept as legitimate the idea that strategic history is unending, just as it had no confidently identifiable beginning. What the concept of the stream of time can do for us is provide all important temporal context. While on the one hand, it accommodates considerable change in the character of strategic affairs, on the other, it insists upon our appreciation of the many, indeed fundamental, continuities in our history. On grasping the meaning of the concept of the great stream of time, scholars, soldiers, and politicians should feel rather less lonely and perhaps afraid in their unavoidable ignorance of the future.

There can be no hiding from a fundamental problem that must drive this discussion; specifically, how can we prepare for our future security when, all too literally, we neither have, nor are able to obtain, a thoroughly reliable understanding of our society's security needs in the future? Restated in short form, how do we go about our defense planning when we cannot know what perils the future will bring? It is appropriate, if more than a little worrisome, to recognize as reality the enduring contextual truth that defense planning must and will be done, no matter the state of our grasp of current, let alone future, dangers.[5]

Bearing in mind the certain fact that our future strategic history can be played out only once regardless of the imaginative alternatives that futurists can devise, it should be clearly understood that history, in common with strategy, is all about consequences that will not usually be easy to understand with well-founded confidence.

Thus far I have chosen to emphasize the scale and the quality of the problem that must confront defense planners. The basic reason for such negativity does not lie in deep-seated pessimism on this analyst's part, but rather in a determination to insist upon full realization of the nature of the challenge to policy and therefore to our strategy. How can we "define the future threat" when that future is a book closed to us forever? It is ironic that, although we are, and will remain, entirely ignorant of the future in detail, by contrast with that bleak ignorance we can, and should, be thoroughly knowledgeable about the context of threat in our society's future. How and why is this remarkable contrast possible? My citation of Thucydides' great *History of the Peloponnesian War*, when considered in tandem with the concept of an unending great stream of time, can provide the basis for an adequate answer.[6] The particular merit for this analysis in Thucydides' history has been well-flagged by many scholars, as well as political and military professionals.[7] However, the core of the lasting value in the Athenian general author's literary historical work has been well expressed succinctly by Robert Gilpin. This American professor has claimed that "The classic history of Thucydides is as meaningful a guide to the behavior of states today [1981] as when it was written in the fifth century B.C."[8]

Gilpin also offers the following thought that many of our contemporary scholars most probably would deem close to absurd, were they ever prepared to consider it seriously.

> But, in honesty, one must inquire whether or not twentieth-century students of international relations know anything that Thucydides and his fifth-century compatriots did not know about the behavior of states. What advice could today's students give that would have enabled the Greeks to have prevented the great war that destroyed their civilization?[9]

We know for certain that Thucydides believed that his then contemporary history of the great war that all but destroyed Greek civilization was, in effect, the story of the persisting, indeed permanent, history of mankind. If we find this belief plausible, at the very least, then it means that we have valuable, if admittedly variable, access to an abundance of evidence with high educational merit for our defense planners today. We do not and cannot know the strategic future in detail, but we should have more or less available to us the understanding of what was probably considered and actually attempted, over the course of 2 1/2 millennia of global strategic history. Obviously, very large questions about change and continuity intrude aggressively. How do we cope intellectually and practicably with the cumulatively gigantic changes that unquestionably have occurred in strategic history?

It is necessary to begin with a treatment of the most basic questions that can be posed. Probably the most fundamental of issues pertains to the distinction that one might wish to make between continuity and change. Directly posed for the sake of this analysis, just how similar in kind are critically important happenings in 5th century B.C. Greece to the events in our world in the 21st century? We are apt to talk rather freely, perhaps even glibly, about the politics, policy, strategy, and tactics of then and now, even as if the huge differences in thought, behavior, and material either did not happen or, at most, are relatively unimportant. Undeniably, a peril of possible historical anachronism must loom menacingly over this discussion.[10] It is my contention that although material, and therefore necessarily tactical, change has been continuous, if irregular in pace, over the course of strategic history; the larger challenges have remained all but constant. There has been, and there continues to be, some triangular tensions among technological change, tactical ideas on the best way to exploit new capabilities, and the cultural preference, assumptions, and expectations that often seem to rule important aspects of our strategic security.

The reader of Thucydides today can hardly help but notice how familiar the dilemmas and anxieties of his world appear to us. It is prudent to be skeptical of an easy assumption of likeness between his political and strategic world and ours, but, alas, it is even easier to be dismissive summarily of apparent similarities. It is a considerable aid to understanding if one is able to probe for the motive and method behind strategic behavior. Such a forensic endeavor is rendered more feasible than it might be, certainly intellectually more legitimate, if one is prepared to accept as a working hypothesis the large notion already introduced here of there being a continuous great stream of time. Once the awful and awesome Peloponnesian War is placed where it properly belongs, as a dreadful episode in strategic history, we can see our own experience in strategic history as helpful in longer-term temporal context. What is contemporary for us actually is neither more nor less authentic and distinctive an experience than were the sundry alarms and traumas of the ancient Greeks. Of course, threats and dangers always are specific to particular circumstances historically, but there is high value in being able, from time to time, to elevate one's gaze above the "threat of the week" and the alarm of this afternoon in order to attempt to secure a better understanding of the structure and working of the process that produces estimates of future menace. Behind the central working hypothesis of this analysis, which affirms the strength of strategic historical continuity, is a robustly empirical theory that explains the more fundamental reasons for that enduring reality.

To understand future threat, it should be realized that the 2 1/2 millennia of strategic history fairly accessible to us can and should be utilized in order to generate some theory with explanatory power, at least potential, over the rich and characteristically ever-changing flow of events. Fortunately, we do have enough to hand some grip and grasp on the principal factors that, in combination, often malign and drive our strategic history.[11] Specifically, strategic history can be approached and understood as the ever dynamic outcome of relations among human nature, political process, and strategic logic and method. It is my argument that none of these three broad driving forces in history are discretionary. As human beings, we are what we are and, effectively, always have been. Cultures have come and gone, but the human nature that adopted and adapted them has been all but constant.

My second factor, political process, has been manifested in a wide range of forms, but we ought not to be misled into believing that our species is capable of achieving such a perfection of political form that all questions about threat would become strictly moot. Political process is universal because the human condition—Man's Estate—is ever perceived either to be under threat, or at least likely to be so. This intimation of menace, in a near or far term, requires making decisions in the interest of alleviating anxiety or actually meeting menace. The process of providing for community security is what we understand as politics. To provide security, there has to be a political process.[12] The details of this process do not really matter for the integrity of this argument. Political process determines who is relatively the more influential among the concerned citizens, and the more or less well-established facts of the local political process confer the necessary legitimacy.[13] Of course, beliefs and values have varied immensely over the whole wide range of human security communities. But, the ubiquity and necessity for legitimating political process has been a reality throughout all strategic history.

The third and final factor in my austerely economical theory of strategic history is the universal necessity for strategic logic and method. By strategic logic and method, I mean the approach to challenges that endeavors to protect or secure desirable political Ends, employing effective Ways, through persuasive application of generally coercive Means.[14] The basic triad of Ends, Ways, and Means needs appreciation in the context of the prevailing assumptions, meaning beliefs that may or may not be well-founded factually. The strategy triad can look as neat and generally efficient in its glorious inclusivity, particularly in PowerPoint form, as actually it is likely to mislead the unwary. The basic structure of the theory of strategy is as appealing in its wonderful simplicity as it is certain to frustrate in attempted practice. No matter how persuasive the core triadic theory of strategy appears to be, the military practitioner will long remain in a condition of ignorance as to the depth and scope of the difficult challenges he needs to know how to meet.[15] To cite only the most obvious of practical challenges that an unduly cursory mastery of the PowerPoint presentation of the theory of strategy most likely fails to register very clearly:

1. Policy **Ends**: These are selected as a result of political process, not necessarily on the basis of strategic feasibility. These Ends almost always are regarded by their political parents as "work in progress."

2. Strategic **Ways**: Because strategy is, in its nature, an adversarial activity, there will rarely be ways that can be presumed reliably ahead of time to be effective.

3. Military **Means**: Military effect, let alone strategic and political effect, is not definitively predictable. Estimation of anticipated nominal and notional capabilities may prove not to offer reliable guidance for the future. Strategists sometimes need to be reminded of the eternal authority of tactical performance.[16] While strategy is needed to ensure that tactics are politically useful, it is well to remember that all strategy, everywhere and at all times, has to be done tactically. Strictly understood, there is no action at the strategic level of war, common misuse of "strategic" as an adjective, notwithstanding (e.g., there are no "strategic troops," because all troops behave with strategic effect and therefore logically should be understood to be "strategic").

4. **Assumptions**: Because the future, by its nature, denies us reliable knowledge, the defense planner has no option but to assume that many, perhaps most, of his beliefs will prove to be true enough. Scientific, which is to say empirically repeatedly testable, under-

standing of the future security context literally is unobtainable. This means that much of a state's defense planning is obliged to proceed on the basis of anticipation, hope, expectation, and possibly confident beliefs about the future—which cannot be tested for their accuracy. It is as frustrating as it is important to recognize that tomorrow cannot come on human demand; it is always a future wherein currently unexpected alternatives may occur.

It is necessary, but by no means in itself satisfactory, to endorse Raymond Aron's famous praise of prudence as constituting the primary virtue in statecraft.[17] It is no simple matter to behave prudently. It is undoubtedly highly desirable for defense planners to be respectful of prudence as a value, even as the most important value, but further thought reveals that prudence does not come accompanied by advice as to behavior reliably applicable to it. It is indeed wise to appreciate the relative significance of the consequences of thought and actions, but, unfortunately, the need to anticipate consequences requires some capability in the field of anticipation and even prediction, which is why this monograph now turns directly to confront the challenge of possible future threat. How can we endeavor to deal with the uncomfortable fact that we must cope with an unknown and indeed significantly unknowable country—the future?

CONTEXT

The first Bill Clinton administration found itself more than marginally embarrassed by the condition of unchallenged international strategic preeminence, which it inherited from the relatively exciting and successful George H. W. Bush years. Strategic preeminence had been an aspiration in the Jimmy Carter and, even more, the Ronald Reagan years, but it would be a gross exaggeration to try to claim that it was seriously anticipated, let alone predicted and expected. Far from being self-congratulatory at the apparent arrival of a somewhat unexpected hegemony, the United States and its defense community gave a convincing public impression of notable bemusement, bafflement even, by achievement of unchallenged and unexpected unipolarity. The country had mobilized in the diverse ways of grand strategy in order to see off a Soviet menace that had appeared to be gaining in strength through the 1970s. Under grand strategic pressure, however, the Soviet challenge first weakened and then nearly vanished during the Bush presidency. The United States and its principal allies were counting the actual and believed imminent strategic rewards that should flow from meeting satisfactorily. Soviet challenge that had been burgeoning since the late-1960s died politically, or at least gave almost every appearance of doing so. Of course, there were some prudent residual American doubts as to the permanence of the change, transformation even, in Moscow, but as the 1990s proceeded, it was unquestionable that the Soviet demise was a genuine and apparently lasting collapse and retreat from imperium.

Although it is exhilarating when the major, indeed the only premier-league, menace to national security suddenly disappears, it is more than slightly disconcerting and even baffling. The fact that it has happened before—think of 1865, 1919, and 1945-47—can offer scant consolation. For more than 40 years, the American defense establishment had grappled with the certain knowledge that the Soviet Union was by far in a league of its own—the only serious danger for the Republic. One, arguably two, whole generations of American officials and Soldiers had understood that, ultimately, the United States had only a single enemy with which to contend, arguably worthy of respect as a near, certainly self-intending, strategic peer: the Soviet Union.

For understandable reasons of its geography and history, the United States is ill-prepared to cope with the problems for national security in these mid-2010s and beyond. Of course, the evidence to be provided by contemporary history cannot be trusted to suggest how we should proceed in our planning and deployments for national and international security. Only rarely in the country's history is there need for a debate on the basic objectives, and the necessary strategy to meet what has become, or plainly is becoming, a new situation. It is obvious, for example, that this monograph must pause in order to address the evermore pressing issues of national security context. It would make little sense to attempt to meet the obvious challenge posed in the title here—"defining the future threat"—unless one first can establish just what it is that Americans care about that may be threatened, and indeed why and how much they should care.

I must confess to surrender to the authority of the notional thought that strategic reason and logic are intimately engaged with this subject. I assume that threats in the future will be no more random than will be our organized efforts to thwart them. It is necessary not to flinch from risking an apparently exceedingly elementary level of inquiry here. It should be obvious that, because I am thinking strategically, I am assuming the authority of an international adversarial context. Lurking in this discussion is appreciation of a fundamental element necessary for such a context. By and large, it is sensible to consider the austere theory of strategy with its Ends, Ways, and Means in a context of relationships. To clarify: each of the three organizing elements in the traditional formula of and for theory requires either a plain notional or more likely some actual physical and ideational opposition. My immediate destination here is to "pause" or perhaps "stop" the definitive political and strategic end to the Cold War that menaced our security for more than 40 years.

Twenty years ago, many people believed that the United States had a "unipolar moment" of unpredictable duration.[18] Relatively minor distraction in the Balkans and in the Middle East continued to frustrate U.S. policy and strategy, but such annoyance was not judged terribly serious by Americans. Jihadist terrorists emerged from the rubble of erstwhile Soviet ravaged Afghanistan, though more to be noted than really noticed as objects of menace requiring urgent political and strategic attention. The American attitude in the 1990s amounted in potential strategic practice to the view that the sole remaining superpower does not clean windows just because there are no more demanding missions on offer. The all too practical problem for the U.S. defense community in the 1990s was a serious absence of political motivation. In political effect, in the 1990s, the United States found itself deprived of foreign threat, a condition last enjoyed in the early-1930s (though I could register the fact of some American hostility toward the empire of Japan following that country's aggression over Manchuria in 1931). While defense professionals will never be caught entirely naked of threat perceptions and concerns, such is not the case for the civilian public or, in good part as an understandable and appropriate consequence, for that public's elected representatives at every level of political responsibility. This is popular democracy in action, and it was as unmistakable a phenomenon in the Athens of the 5th century B.C. as it proved in our 1990s.[19] When an electorate is far less frightened than it had been recently, it will expect and, if politically necessary, demand that public expenditure on military defense preparation be reduced markedly. In the 1990s, overwhelming foreign danger, duly conceived and not infrequently presented as a threat, seemed simply to disappear. Hardened defense professionals failed to celebrate unduly but not because their personal livelihoods were at risk. The reason for moderation in celebration by defense experts lay in their conviction that strategic challenge had not been defeated definitively, but rather was only resting while new contextual fuel for conflict was amassed by troublemakers, familiar and novel.

In a universe characterized more by prudential reasoning than is ours, one could have expected a "1990s" decade to be employed as a period of strategic pause for the purpose, *inter alia,* of using the luxury of the relaxation of serious security concern, let alone alarm, to revisit the fundamental assumptions of American grand strategy.[20] One needs to bear in mind the fact that the country had been fixed in a particular character of response mode for more than 40 years by the Soviet threat in its several forms. However, democracies tend not to interpret the absence of acute foreign danger simply as a "time out" from conflict as usual, but rather as apparently unmistakable evidence of inevitable political progress. Politicians know that they will not receive support if their message is one electors do not like. The truth of the matter was revealed for all subsequent time by Thucydides, of course. He knew that history always is punctuated by episodes marked by appalling errors in human judgment and by chance, as well as by periods characterized mainly by prudence in the conduct of public affairs. Proper grasp of the implications of the concept of the great (and unending and in fundamental effect) stream of time can and should protect us both from excessive alarm as well as naïve faith that some contemporary evidence of political wisdom offers unmistakable evidence of the approach of the long awaited "End of History." Comprehension that history has no end (we hope, thinking of residual nuclear concern) also must carry with it the understanding that serious national security concerns may only episodically be alarming, but certainly it will feature repeatedly in the future. In short, the United States does not and will not face security problems that, once coped with adequately, must be followed by an endless ever improving period of peace and prosperity. The political and moral context for this discussion has to be assumed to be one of eternal periodic alarm in varying degree. This is the permanent reality of all strategic history, the American included. It is, and must remain, a strategic history and is not to be confused with a fairy tale that climaxes with a definitively happy ending.

As a matter of empirical record, the decade of the 1990s that should have been characterized as a strategic pause that allowed for a reconsideration of national grand strategy instead was a decade marked by technical and largely tactical obsession with the promise that there might be in digital information technology. For the better part of 10 years, in effect, the American defense community all but squandered its inevitably temporary unipolar moment, self-obsessively contemplating how tactically formidable it might become as a result of digitization.[21] The revolution in military affairs (RMA), military revolution, and military technical revolution collectively were treated as being the strategic flavor of the decade. The inattention to more serious matters of policy and strategy was revealed in painful form in the political and strategic errors that succeeded September 11, 2001 in the counterinsurgency (COIN) decade of the 2000s. Digitization in its many applications has proved, unsurprisingly, to be of high tactical importance. The problem cited in this section on context essentially has nothing to do with the ever greater technical and tactical exploitation of information technology, but rather with the subjects of yet higher professional concern to us that have been neglected unduly in the fields of strategy and politics. In noteworthy part, the neglect was, *seriatim*, of obsession with the exciting technical and tactical promise in RMA, and then with the challenge of attempting to counter terrorism and insurgency in distant and culturally ill-understood lands.

The better historians among us have tried to advise that policy and strategy inherently are relatively more important than are technology and tactics.[22] This does not amount to the claim that tactics are unimportant, only to the proposition that strategic history has demonstrated that, while tactical and much technical error proved correctible and survivable, almost invariably, political and strategic mistakes were fatal, no matter how worthy the ultimate political and moral intentions may have been.

SPECTRUM OF CONCERN

Unless there is unmistakably good evidence suggesting the probability of particular threat in the future, it is necessary to recognize the fact and impressive degree of uncertainty that is in the nature of this crucial subject.

Capability and Intention.

Threat should be defined as an intention to do harm that logically and practicably requires command of the physical ability. As a general rule, threats have to be received clearly enough by their intended addressees. However, both common sense and strategic historical experience suggest that the unambiguous delivery and receipt of threats also are necessary if they are to perform the function intended by their authors. More to the relevant point and as the theory behind nuclear strategy long has sought to explain, the credibility of some threats almost necessarily will long be in doubt.[23] It is important not to forget that threat manufacture, transmission, receipt, and interpretation is always an essentially human enterprise. The deadly (nuclear-armed) machines of mass menace that compass the crown jewels of potential threat projection will be inert instruments that function awesomely only as a result of the outcome(s) to human adversarial behavior. This whole discussion of threat and responses to it is suffused with recognition of the salience on both sides of distinctively human thought and behavior.[24] Nearly all military capability lends itself to some potential mission of coercion, though most of the time and in most places inherent ambiguity about motives for employment simply remains as a strategically valuable residual quality of military power.

It is usual to define threat with usefully pointed economy as being well enough expressed by the formula: threat = capability x political intention. There is virtue in this austere expression, but little additional thought is required in order to understand that political intention does not usually remain constant for long. In other words, this simple formula may well mislead us into regarding as constant what in reality usually is anything but. Without wishing to divert this analysis away from its central theme of "threat," nonetheless, it matters deeply that readers understand fully that, in strategic history, most of the time, policy Ends are not fixed permanently. Because conflict and war are, in practice, kin to organic happenings that wax and wane often unexpectedly as they develop, so policy Ends quite typically will need to be adjusted in order to reflect a clearer more contemporary understanding of what appears to be possible.[25]

To extend the point just registered about usual fluctuations that characterize policy and its Ends, it is necessary to accept that this whole subject is liable to real-time adjustment, even on items widely held to be of fundamental significance. Perhaps the easiest way to explain what I mean is to contrast the austere formal architecture of strategy with the reality of strategic behavior in historical practice. Whereas one can teach that strategy entails the pursuit of politically agreed policy Ends, by suitable strategic Ways, employing the necessary military Means, students also need to have a secure grasp on how theory actually is, and has always been, practiced. In reality, each basic element in the simple strategy formula—Ends, Ways, and Means—is ever likely to be in need of adjustment for a better fit with changing circumstances and adaptation in vital detail.[26] The argument was made earlier that the "Ends" of policy should always be regarded as potentially being only a "work in progress" as contrasted with some conclusive vision of intention. In fact, the same tolerance of some flexibility is necessary in regard to each of the big three categorizing concepts—Ends, Ways, and Means. Governing this whole discussion is the eternal and ubiquitous truth that statecraft and strategy are in the realm of what proves possible to achieve, not necessarily of what ideally it may be highly desirable to accomplish. Strategic history demonstrates clearly that defense planning as a pragmatically prudent matter always requires us to be satisfied with "good enough" solutions.[27]

The professional soldier typically favors precision and consistency on the part of political superiors. This liking for a clear, definite, and preferably unchanging strategy cannot guarantee delivery of a steady and well-directed military effort, but it should help.[28] The politician who is a policymaker is more likely to regard politics and political effect as the home of Ends, rather than the policy that he hopes will serve adequately to support the politics. In other words, while the soldier understandably tends to view policy objectives as his relevant Ends, with political process holding on a secondary role, the policymaker typically regards politics and its process as the true authority, albeit one expressed and explained in commonly (but permanently) temporary policy terms. The better definitions of strategy take care to specify, at least to suggest, the superior role played by political process in the making and execution of strategy. My own preferred definition risks erring on the side of explicit recognition of the relative but clearly superior importance of politics.

I choose to define military strategy as "the direction and use made of force and the threat of force for the purposes of policy as decided by politics." The context for political activity and decision is fundamentally different from that for military decision. It follows necessarily, though commonly unappreciated, that the understanding of future threat also is likely to diverge as between the different worldviews of the professions of politician and soldier. Of course, threat definition requires military analysis, but it is necessary to remember that soldiers are not usually in charge of a country's statecraft. Statesmen are politicians who have succeeded in playing the game of relative political influence more successfully than have their rivals.

An important migration in conceptual meaning concerning strategy occurred in the 19th century, but it is not as well understood as it needs to be. Specifically, the core meaning of strategy in the 1820s was that its purpose was to direct choice of and for battle in its expected consequences in war. In noteworthy contrast, strategy came to be regarded post-World War I primarily as a support for politics and its policy in peacetime and wartime. Modern strategy has not lost its erstwhile battle focus, but that fixation of Clausewitz's day has been conclusively widened in political scope.

As often as not, there will be a vital quality of ambiguity about actions and words intended to carry menace. Professional soldiers favor, perhaps need, definite orders and commands, the dangers associated with it which can be calibrated by some military calculation. But, the politician, in his characteristic universe, wants to be able to advance further or to retreat with minimum political embarrassment. Indeed, the political leader can be endangered by self-deception that may confuse his strategy. Because they act preeminently with and through words, politicians are prone to misread the scope of feasibility in policy zones that transcend the world of politics. This danger understandably is especially acute in the relations between politics and war. In fact, although all war is political in meaning for all time, it is not an integral branch of politics. Although war should serve politics, war is not politics, notwithstanding the migration in meaning of strategy noted previously. There is a permanent need for currency conversion between military threat and action and desired political effect. Although threat to American security can be so obvious as to be all but unarguable, that reliable, if still uncomfortable, relative stability in assumption about foreign menace is by no means dominant in our contemporary threat analysis and divination.

There is little room for doubt that Thucydides' understanding of the causes of the great war of his time between Athens and Sparta can offer guidance superior to the alternatives from our contemporary social science.[29] In political life, both domestic and international motivation inevitably is always complex and therefore potentially complicated and ultimately indeterminate if one looks foolishly for a scientific quality of evidence. But, if we choose to be content with high plausibility of fit with what is known with substantial confidence about political context, it is possible to identify probable motivation of the causation of events. It should be recalled that threat has been defined here as a product of capability multiplied by intention. In other words and by way of example, a force of intercontinental ballistic missiles is naked of any specific strategic meaning until its political owners decide upon such.

It is worth noting that, over the past century, many scholars and politicians who should have known better gave robust indication of their failure to grasp the essential point just registered here. The whole modern history of arms control has revealed confusion of understanding about the significance of arms in their relation to political intentions. Identity of political ownership of weapons largely, though not absolutely invariably, is key to understanding strategic and political meaning. Military capability may well be rich in strategic, operational, and tactical implications, but the ascription of threat depends upon the political ownership of the instruments of interest. Of course, such ownership often will be innocent of malign intention, or at the least will only be deemed likely to be contingently menacing.

Since context typically drives contingency, and given that context should lend itself to influence by behavior that shapes political judgment, the grim possibilities that one can identify with particular inert military items may serve as providing timely warning for statecraft. Episodically throughout recorded strategic history, developments have been interpreted as being in an adversarial context, and the identification, possibly misidentification, of great security threats has ensued.

Thucydides was a deep and subtle analyst of the meaning of his times, and he expressed little doubt about the primary cause of the great war that was the ruin of ancient Greece. The Athenian historian (and failed general) acknowledged the many complaints that Athens and Sparta lodged against each other, often with some justice, but he declined admirably to be distracted even by particulars, admittedly contributing to the main thrust of his narrative and the clarity of explanation that he sought to convey. In words alive with meaning for all subsequent periods, especially for ours in the 21st century, Thucydides wrote as follows:

> To the question why [in 431 B.C.] they broke the treaty [the Thirty Years' Peace of 446 B.C.], I answer by placing first an account of their grounds of complaint and points of difference, that no one may ever have to ask the immediate cause which plunged the Hellenes into a war of such magnitude. The real cause, however, I consider to be the one which was formally most kept out of sight. The growth of the power of Athens, and the alarm which this inspired in Sparta, made war inevitable. Still it is well to give the grounds alleged by either side, which led to the dissolution of the treaty and the breaking out of the war.[30]

The Athenian Empire demonstrated an intolerance of political and strategic dissent from its client allies and certainly inadvertently provided ample fuel for hostility. However, as Thucydides recognized and acknowledged, the security problems for Athens in the longer term was fundamentally the immoderateness of her power as perceived by other Greeks, and then all too naturally the unmistakable evidence of an overweening confidence evidenced in Athenian policy and strategy as an easily traceable consequence. But, what is the relevance of the Athenian case to the American defining of future threat? Indeed, is there any relevance worthy of much note? The most pertinent problems for American security need to be understood as political, though there is a major, perhaps even overwhelming, systemic complication. Through all of strategic history to date, the principal and indeed enduring threat to security has lurked in unbalanced power. Balancing power cannot guarantee peace and tolerable prosperity, but we do know that the evidence proved by unmistakable historical cases of unbalanced power, or certainly the fear of such, inherently are hostile to that peace and prosperity.

Unfortunately, the Thucydidean worldview, which was and remains all too realistic, tells us that our future security is certain to be threatened by the usual, indeed normal, working of the politics of international life. The distinctive advantages and benefits that America enjoys are collectively the motive that will find expression in future threat. This is not quite to suggest that the sense of hegemonic preeminence and duty that led the United States to endeavor to sustain a non-Communist South Vietnam in the 1960s, and a generation later to redirect and re-make Middle Eastern and then Central Asian history in Iraq and Afghanistan, was unworthy. But, it is to suggest that it was unwise. Prudence in policy is much easier to locate in historical retrospect than it was at the time. Frequently in strategic history, good intentions have been permitted to masquerade as prudence. For the record, even near-term achievable political and strategic results cannot necessarily be trusted to deserve judgment as prudent.

Risk and Danger.

Virtue and villainy do not bring their most appropriate consequences inevitably, but it is prudent to assume that such might be the case. Notwithstanding its continuing relative greatness as a superpower, questions fundamental to this discussion have to include the following: What will be the political context for America's definition of threat in the future; and what might America strive to do to exercise some helpful measure of influence in, perhaps even control over, that future political context?

There is good reason why time plainly has to be understood as comprising the least forgiving of strategy's many dimensions. The temporal factor is inalienable as a critical element in all defense planning. Although this has always been true throughout strategic history, never has it been more vitally salient than today. Defense planning must be geared for tolerable fit with the anticipated timing of foreign security risk and danger. Indeed, it should be all but second nature for planners to be able to categorize menaces to national security on a time scale from current alarm through near- and medium-term worry and serious concern, to the more agreeable distant horizon of anxieties about the relatively far-term future. Future security threats are anticipatable, let alone predictable, only with diminishing confidence, as one ventures necessarily and somewhat imaginatively into an evermore distant future.

Moreover, if particular political antagonism and hostility is less than confidently predictable, it is difficult to specify possible or probable measurements attached to the strategic future. When the political context changes, so must the strategic one. Given that strategic meaning is conferred by political effort, it is obvious that exercises in military-technological futurology are likely to miss most of the needed argument concerning prudent defense provision. However, contrary perhaps to a line of reasoning that would privilege an undisciplined openness of strategic competition, this discussion is sustained by the conviction that there is and will remain what can be regarded as a categorical imperative shaping of international political and strategic competition. Without quite endorsing a fully deterministic view of the terms for the future security of America, nonetheless there are substantial grounds for confident identification of principal strategic historical markers of the policy passage that need to be pursued in the interest of prudence. It is advisable that I remind readers of the genuine uncertainties and possible challenges that our defense planners have to recognize and work around, if not overcome. At some risk of encouraging pessimism, I would not endorse seeking to define future threat. The United States has to be able to cope with the following uncertainties:

- Identity of menacers (or possibly threateners)
- Seriousness of menace
- Time horizons of menace
- Feasibility of adequate response (e.g., time, cost, likely consequences).

Contemporary uncertainties must preclude the kind of exercise in threat definition that can produce all but tangible and convincing answers. It must follow, therefore, that the relevant challenge to the U.S. Army should not primarily be one of threat predictions and spotting! The reason why Thucydides' great book figures so significantly here is not because I am striving to assert the relevance of an ancient Greek historical analogy for today and tomorrow, but rather because the context of competition concerning security is so categorically analogous. The differences between 5th century B.C. and today are indeed enormous, but so are the similarities. Aside from the quality of catastrophe that we must assume would follow nuclear use, there is little about the relations between Athens and Sparta that is notably mysterious to us today. However, what matters most for the subject of this discussion is the plausibility that the political and strategic relations among the greater powers have not altered fundamentally since that now distant period. This is a vitally important subject about which to theorize and argue, because it bears with maximum directness upon the issue of whether or not strategic history in the great stream of time can provide sufficient evidence of empirical examples of interstate behavior, prudent or foolish.

The study of history shows us what can happen, which we know for certain because it has happened. It is not available as a reliable guide to accurate prediction, because the course of strategic history is subject to too many and too diverse a range of possible influences, not least the ever potentially individual human, to provide adequate fuel for confident prediction.[31] The United States should not seek, because it would not find, specific instructions in Thucydides as to what should be defined as threatening. But, what it can and should find is education both on how great powers behave and what the hazards may prove to be in that behavior. The definition of future threat should encourage prudential defense planning in line with the country's desired, albeit self-disciplined, role and responsibility in the international system.

Risk and danger cannot be eschewed and thereby avoided conclusively, but the whole record of strategic history offers education in the high value of balancing power. I must add hastily that history also warns of the practical difficulty in achieving such balance, and especially of the risks and danger that are apt to beset the pursuit of immoderate greatness. Preeminent unilateral hegemony will always be challengeable sooner or later, typically the former, but it need not be effected in a power transition process fatal for world order. Power transition necessarily has characterized all of strategic history, but to date, of course, it has never been coerced or negotiated in the context of nuclear armament. American preeminence as a, and then the only, superpower for a while in the 1990s is particularly peril-prone both because of its nuclear backstop and the reluctance or even inability of many Americans to see themselves as they need to in the great stream of human history. U.S. superstate prominence has been a physical and psychological reality since 1943–44, but this reality requires understanding as a passing dynamic episode contextual in history.

When, as here, we think about future threat, we have to consider how the United States may best continue to perform a benign hegemonic role in international security. There is a problem in that Americans have difficulty understanding that their country, notwithstanding its unique features, in the last analysis simply is yet another state that has no practicable choice other than to play the game of nations along with everyone else. There is only one game, and it has proceeded through all of strategic history from the time of Athenian and Spartan competitive preeminence until today. Contemplating the future threat to security, we should proceed unconfused by our domestic ideology of uniqueness. Future threat will lurk in the malign influence of foreigners' fear, their energetic determination to protect their reputations, and in their definition of national interests that may not be compatible with our own.

Thucydides wrote about the great war of his lifetime, on the safe assumption that there would be other great wars in other times involving other polities. He was unsparing of human folly and error, and he believed that his revelation of such might have some useful benign educational effect. Following the Greek author in spirit at least, I believe that a robustly and sufficiently evidenced grasp of how states need to behave in order to balance power is key to preserving the international order vital to future security. Ironically perhaps, America's ability to define future threat in a way conducive to the protection of important values is hindered by the experience and legacy of extraordinary national hegemony. This is understandable, even if not well understood currently. An important source of difficulty lies in the relative modernity of active American participation in the game of nations. The genuine risks and dangers of the Cold War served in practice to hide the fact that the great Soviet–American Cold War was far from unprecedented in strategic history, save only for the uniquely challenging addition of weapons of mass destruction to the equation of high risk.

It would be misleading to characterize effort for the future threat definition as an attempt to return international politics to a more normal multipolar condition, following the decades of bipolar Cold War, the unipolarity of the 1990s, and the unprofitable, if well-intentioned endeavors to remake backward realms in the anti-jihadist 2000s. But, given that the United States was not a very serious player in international politics prior to World War II, it is scarcely surprising that the country has relatively little understanding of how and why international politics are played as they are. Contrary to appearances, this is not meant to be critical. American statecraft culturally is what it is, reflecting somewhat the novelty of the national experience. This discussion does not seek to engage combatively with an American culture that reflects somewhat the dominant national experience.[32] The challenge here is to explore the best way the United States, with its distinctive and quite recent historical experience, can cope with the challenge of defining future threat.

A key reason why it is advisable to accept the concept of prudence as having merit superior to alternatives is because it is the idea that most exactly captures the critical difference between a risk and a danger. Moreover, prudence is a concept that, in theory at least, suggests in its required focus upon consequences how potential dilemmas should be approached and met. Unfortunately, causes for concern in international politics do not appear on policymakers' desks helpfully labelled either as risky or dangerous, while their actuality or their potentiality for harm may well encourage uncomfortably subjective and unmistakably uncertain threat assessment. It is prudent to be clear in one's own mind about the distinction between risk and danger, albeit while ascertaining the necessarily subjective character of the judgment offered. Logically, a risk is a potential way station that can be en route to another condition, which one may well judge to be dangerous. Given that danger is a realized condition, accurately perceived or not, risk has to be regarded as a quality contributing to the maturing of incipient peril. Plainly, the ascending peril between risk and danger is captured adequately in the language used. Where the idea of prudence intrudes most usefully is in its ability to suggest persuasively that there is a vitally important difference between the two. Poor statecraft and strategy, unaided by the insight required by and for prudence, will be prone to conflate risks and danger, with the result that we may miss important possibilities for the control of escalation in menace.

Because of its significantly open nature, the concept of the future lends itself easily to abuse by theorists who are not careful in their choice of words, either because of what may be attributed to mischievous aforethought or, more likely, simply because they are theorizing beyond their normal comfortable intellectual depth. One reason we should seek to discipline our choice in conceptual usage is because we do not know the best way to influence the future: at the very least, we should take care to avoid misusing concepts that might do unanticipated damage to our security. The problem most central to this discussion is the unavoidable fact that we are unable to foresee the future. When our intelligence community looks forward in time, it is trying to identify the consequences of past, current, and future historical phenomena. A determination to be alert to consequences translates as a commitment to prudence. Such a commitment does not necessarily mean the attempted eschewal of risk, including ones that could be dangerous on any reasonable assessment. Instead, what would be meant, certainly implied, is that the possible consequences of our behavior, positive and negative, would be considered seriously. Given the unavoidably speculative nature of all futurology, it is always more or less difficult to ensure that political and strategic decisions about the future that are popular are subjected to honest assay.

The defining of future threat is a behavior exceptionally inviting to the ill-deserved authority of hope over realistic expectation. This discussion now must address key issues pertaining to assumptions and evidence that should not be permitted to slide out of sight inadequately treated.

When Is a Threat Not a Threat? Assumptions and Evidence.

Probably the soundest judgment on the general problem of threat definition for the future was drafted a generation ago by Professor Klaus Knorr of Princeton University. The professor advised with these well-considered words:

> The act of threat perception creates an image of reality; it is a device, a hypothesis. Indeed, this holds true of all perception. All human awareness is a personal construct, something that we organize. It is a set of assumptions that deals with the outside world selectively, focusing on some components while screening out others. We continuously "bet" about the nature of reality. Preexisting assumptions (i.e., theories) about the outside world help us to select our "bet." But for this very reason, they hinder, as well as help, perception.[33]

The other intellectual difficulty vis-à-vis threat perception is that it concerns the future, and often not just the immediate future. There is no information whatever about the future. The political and social phenomena that are objects of threat perceptions change over time. Even if we know what is true today, we could not know what will be true tomorrow. As historical accounts demonstrate again and again, state behavior is highly unpredictable. It can change suddenly; surprise is frequent and not rarely dramatic. All we can do is speculate on the shape of future events by studying the relevant (especially the recent) past. The mere projection into the future of what we have learned about the past already has been pointed out, is extremely hazardous.

Defining future threat is a speculative enterprise that invariably requires the making of decisions on our part that quite literally cannot be justified with reference to hard and definite evidence. For the defense planner, the future is a foreign country overpopulated with situational possibilities abundantly full of threat potential. Because the United States is choosing to remain a (if not the only) global superpower, the threat horizon of the future is well populated. But, we need to ask, just what is a threat and what makes it so? Because the subject lies in the future, there cannot be direct evidence of such threat. What is available, however, is contemporary and recent evidence of behavior and, at least inferable, policy motive, set in the long-term historical context of our understanding of the politics of statecraft. We can know with tolerable reliability what states have done or sought to do recently, and we can speculate with some confidence about the likelihood of contextual continuity in respect of American interests that may be menaced. However, what we cannot do is predict and expect "Black Swan" events or episodes that are, by definition, highly improbable yet potentially and systemically very disruptive in our context of most concern here to the American threat horizon.[34] A Black Swan is an event as unexpected as it proves profound in its consequences. Such occurrences are as startlingly surprising as they are seemingly transformative of prior patterns in near habitual behavior.

It seems unlikely that the outside world will come in haste in the near term to the political rescue of American threat definition with respect to its Army. Menace to particular American interests is locatable worldwide but, both fortunately and unfortunately, currently there is not a single dominant threat filling the horizon. The most prudent conclusion to draw from the current international security context needs to be a distinctly unheroic, but defensibly sensible, determination not to bet extravagantly on any particular category of strategic challenge. The U.S. Army has endured turbulent recent decades of late.

1. In the 1970s, it sought to recover from the diverse but heavy damage suffered in Vietnam.

2. In the 1980s, it was adapting and adjusting both to exciting new technologies (still largely prospective) and to a Soviet threat in Europe that had matured and expanded since the late-1960s.

3. The 1990s saw the Army seeking to make some sense of the highly fashionable RMA and later the transformation possibilities and debates, but the absence from the near horizon of a major dominant military threat considerably undermined the plausible relevance of expert proposals about the technical-tactical frontier and beyond.

4. The 2000s witnessed and appeared to require the return of counterinsurgency. COIN rode again, not least in counterterrorist regard, in the sharpest of contrast to the Army's foci in the two previous post-Vietnam decades.[35]

In summary form, the U.S. Army has experienced decades of: recovery and reform (1970s); modernization for high-end warfare (1980s); fundamental uncertainty and retreat into technical visions of RMA (1990s); and COIN and counterterrorism (2000s). The point that really matters in this discussion is not any particular focus and the political policy and strategy behind it. Rather, it is the fairly obvious lesson that policy, particularity with strategic choices for the Army in the future, cannot be anticipated with high confidence. It would be hard to design strategically more distinctive decades of dominant experience for the U.S. Army than those experienced from the 1960s through the 2000s. The politically dominant strategic reality for the Service was radically different from decade to decade. This is the historical context for consideration of future threat definition. More to the relevant point is the function performed for us by Thucydides in his *Peloponnesian War*, given the transhistorical theory of state behavior that it carries. Thucydides cannot tell us what, or even might, emerge as a threat to security in the remainder of the 21st century, but his understanding and explanation of how and why states behave as they do has profound advisory implications for our contemporary choices in policy and strategy.[36]

Because of operation of the laws of physics, we can secure no grip upon future happenings, no matter how convinced we may be in their anticipation. But what we can know about the future should prove sufficiently serviceable to allow for prudent choices on the shape and composition, size, and equipment of the Army. With much gratitude toward Thucydides, and also as a result of careful study of strategic history, we can identify prudent choices in military posture. This is not a simple matter, because the would-be threat definer, perhaps diviner, needs to be fully alert to the following: the power of accident; strategy's adversarial nature; Black Swan possibilities; enduring human tendencies with conflictual implications (e.g., political process and the implications of fear and anxiety); and the potency of needful ambiguity inherent in much about defense planning and preparation.

It is commonplace, indeed it is only sensible, for defense plans to be specifically contextual to the point of potential political embarrassment. Defense planners as strategists necessarily inhabit a world wherein there are enemies, certainly adversaries, and threat possibilities in the day after tomorrow, if not tomorrow itself (else why conduct the planning?). However, even though threat definition can carry some risk of being self-validating as military professionals simply perform their professional task of preparing for what might happen, such preparation is only prudent. It is important not to forget that the concept, and indeed the principle, of prudence in defense planning means that one must be alert to the dangers that could lie as consequences derivative in part from errors in the plans. Because it makes no sense to seek from the military an understanding of the future that it cannot obtain as a reliable certainty by any known method, we have no practicable alternative other than to settle for an understanding of future strategic need that should be able to rest well enough on our comprehension of the past. We should seek to identify what scholars of strategic history suggest is possible, if not probable. Also, we need to appreciate that, although each period in history may have its own distinctive pattern in events with plausibly presumable motivation, such distinctions are only differences in character, not in nature in the great stream of time.

Learning from History.

History teaches nothing; it is not an agent with motives active on the course of events.[37] But, history is by far the best educator for our strategic future.[38] While we cannot and must not hunt earnestly, though inevitably failingly, for historical analogies to our emerging challenges, we can and should seek assistance, perhaps inspiration, in the intellectual construct known as the "historical parallel."[39] What is claimed using this concept is that the course of events "then" (i.e., whenever) would appear to have shown important features in common with a contemporary track of events of our own recently. To hazard a theatrical comparison, what we would be suggesting would be that much of the state setting appears substantially common to "then and now," while probably also today's adversarial contest appears to have had antecedents in motive and perhaps even in results in past periods. The episodically protracted British strategic experience on the northwest frontier of India is a compelling example of a strategic context that could be regarded plausibly as historically noticeably parallel to the situation in Helmand Province of Afghanistan in the 2000s.[40]

I am careful to distinguish between the analogy and the historical parallel, because the former is fundamentally empirically unsound. History does not repeat itself in ways essential for analogy to deserve authority, but most certainly in contrast, it does repeat itself in the situations that its strategic course poses to policymakers and strategists. It would be difficult to exaggerate the significance of this distinction. To recognize an apparent case of historical parallelism is to notice similarities that may be important between the cases from different historical periods, but that is all. Unfortunately, analysts often latch on to obvious common elements (e.g., Kandahar in the 1880s and the 2000s) between historical episodes in the hope of discovering keys to past success or failure by (inappropriate) analogy. Strategic history simply is too varied in its dynamics and its changing contextual detail for analogy to serve as a valuable source of guidance. There is no alternative to the securing of detailed contemporary contextual knowledge. Old methods cannot be relied upon to continue to be true, if they ever were, just because they may have been well tried in the past.

My intention here is not to condemn or even discourage efforts to seek educational value from understanding the course of strategic history; it is only to advise that such value cannot be found by attempting to analogize in detail from times past. There is no adequate substitute for current knowledge of the details that change. We can and should learn from history about the nature and even the character of future threat. Threat always matures and emerges from causes inherent to the nature of international politics, its statecraft, and its strategy. Thucydides knew and could explain this in the Athens of the 5th century B.C., and his explanation reads convincingly for our era also. What does promote great confusion is the popular tendency to neglect the difference identified here as that which lies between the nature and the character of threat. Probably it would be more accurate to focus on the enduring reasons why threat is an eternal problem, though arguably it is most usefully considered with reference to its nature.

The point of overwhelming significance for this discussion simply is that threat—past, present, and future—is a persistent phenomenon, one consequent upon the enduring need for provision of human security, expressed in political systems that organize for collective protection. Distinctive polities must exist in a whole world of politics, and the relations among them have always been prone to aggravation and anxiety, occasioned by concerns about security. As American defense planners confront the empirically more than marginally mysterious future, they know for certain that anxiety, and periodically even some alarm, about national security will characterize the future. There will be no conclusively happy ending to security fears. The precise character of such alarm cannot currently be anticipated reliably, but the defense community has to be alert to their probable and even possible manifestation, as well as to the causes of anxiety that are actual and current. Important though contemporary details of plausible menace certainly are, it is scarcely less significant to grasp the fact of the certainty of change. Future threat, both empirical as physical hardware and also as foreign intention in our perception, is always in process of some change. The definition of threat changes constantly community-wide. Probably more often than in the physical respect, threat definition shifts as a result of changes in political evaluation.

It may be recalled that much earlier in this monograph, I deployed the familiar bare formula that notionally identified threat = capability x intention. The former typically takes considerable passage of time and notable expenditure of effort to achieve, while the latter can change in a matter of hours. That is not very helpful for would-be scientific appraisal of intensity of threat level, but it is essential to grasp as contextual reality. Notwithstanding some of the aspirations for certainty of knowledge by social science, threat definition and analysis must ever remain an art. Just as governments, let alone key individual statesmen, cannot make vital decisions based upon a true certainty of knowledge of probable consequences, so definition of threat to national security must always ascend, or descend, to being a matter of human judgment.

The strategic history of times past, both far and relatively near, cannot provide a reliable source of evidence as to what should or should not be defined as a future threat to the United States. The reason is because contextual detail must be as significant in the future as it was in the past. It is the beginning of wisdom for defense planners to understand why they ought not seek the literally unknowable. National security planning today cannot sensibly aspire to find its logic on much understanding of, say, the second half of the 2020s or of the 2030s; that might be deemed reliable. But, fortunately, such understanding, especially of particular menaces, is not necessary for American defense planning to be judged tolerably prudent. Unquestionably reliable detail concerning future threat neither exists nor can be generated, given the rarity as well as the scale of the unknown (including the "unknown unknowns").[41] Consequently, our defense community should be satisfied to be guided by an understanding of possibilities, especially of those that current context and recent events indicate as belonging close to what can be categorized as probabilities. This is what should be understood by a prudent approach to national security. Preeminently, prudent definition of threat to American national security has to be an exercise in the analysis of a reasonable projection of the consequences judged probable, or certainly possible, that would follow as a result of serious error in defense planning. To repeat, this is art rather than science.

Probably the definitive terse explanation of state behavior was offered by Thucydides in his rendering of the Athenian reasoning behind its impressive, if ultimately fatal, bid for undue imperial greatness. In the

Athenians' attributed words (of 432 B.C., just prior to the war) provided by Thucydides:

> That [Athenian] empire we acquired not by violence, but because you were unwilling to prosecute to its conclusion the war against the [Persians] barbarians, and because the allies attached themselves to us and spontaneously asked us to assume command. And the nature of the case first compelled us to advance our empire to its present height; fear being our principal motive, though honor and interest afterwards came in.[42]

When we contemplate the international political and strategic context of this 21st century, it is only prudent for us to be alert to the need for a wisdom in national security planning that should be ready enough to cope with challenges to the United States that probably will be ineradicable from the context of contemporary international relations. Even if a Thucydidean fear does not grip the United States, it has to be judged likely that such a dangerous condition of anxiety may well grip Beijing or Moscow, of course, with potentially dangerous consequences for us. Such fear has been lethally, if episodically, characteristic of most international political systems, not least among them our own. It must be a duty of our threat definers to seek to encourage provision of American response that is both adequate in discouragement of hostile initiatives and likely to prove encouraging of moderation abroad. At this present time, we do not know how smoothly a process of power transition between the United States and China will be for either side. In truth, neither a crystal ball nor historical scholarship can be of much assistance to American defense planners today: the strategic future is not knowable in important detail.

However, we do know of the need to be alert to the danger in a major process of international political transition of power. Prudence, as a guiding light, is the basic requirement for our national security in the decades ahead. Threat definition will change in character of detail, but a fundamentally prudent American national security posture should be well able to cope, notwithstanding the uncertainty in anticipation.

CONCLUSIONS AND RECOMMENDATIONS

Prudence and Ignorance.

Because the future is unknown and unknowable in detail, it is the duty of the American military establishment to be as ready to confront and defeat threats as society's contemporary political support allows. The ability to define future threat is not a skill that can be taught. Highly relevant, indeed vitally necessary, data is missing. Prudence is the most useful concept that should be adopted as the idea most fitting for authoritative service in the current era. A prudent defense must be one tolerant of some error. The United States cannot promote its global interests equally regardless of character and intensity of menace. But it can and should be ready to respond to broad categories of security challenge. Just how broad is a matter for political decision superior to that under consideration here as threat definition. It has to be understood that there are two distinctive, albeit related, aspects of the future threat challenge. On the one hand, there is the necessity for Americans to decide what is and what is not a threat, as contrasted with a risk or residual source of danger. On the other hand, it is necessary to understand what might be done about the apparent menace in question.

Because it remains the global superpower, the United States is exceedingly reluctant to appear indifferent to the security anxieties of its clients. The United States is learning today what has been true at all times and in all places in the past, that the sovereign "working level" of threat definition often devolves in undesirable practice upon some of the threat definer's clients in the ranks of the net security provider's foreign dependents. A prudent defense is one that is ready enough to meet threats that have matured or perhaps erupted less than was well anticipated in years previously. The extraordinary geopolitical, and hence geostrategic, scope of America's foreign security provision is the source of exceptional, though certainly not historically unprecedented, anxiety. While identifiable potential dangers great and small need to be covered for potential military intervention, the United States has to be willing and able to attend productively to the structure of security in different regions around the world. The purpose of a prudent defense is to provide sufficient security for whatever may occur. Because precision in the definition of future threat is impossible, it is necessary to limit future ambition to the ability to cope with menaces by broad categories of risk and possible danger. Obviously, it has to follow that American military power needs to be notably mobile and friendly to jointness of enterprise.

Definition of Threat.

There are fortunately very few threats that all but demand to be so interpreted by their political owners. Russian modernization of its strategic nuclear forces and Chinese modernization and large-scale augmentation of its naval assets with obvious access-denial relevance spring to mind as unarguable contemporary examples of military developments abroad that plainly are intended to convey the idea of threat to the United States, at least to America's long-standing interests. However, Americans should recognize that although their country is, in an important sense, simply another great state in strategic history's lengthy procession of such, a combination of technology and historical circumstance has rendered the American role in international security truly exceptional. Imperial Athens was the proud possessor of distinctively unusual greatness in the 5th century B.C., but it discovered painfully that such greatness did not always translate into a usable superiority of power, e.g., in Egypt and later in Sicily.[43] The contrast, but also some similarities, with the American experience from the 1960s until today almost beg for explicit notice.

This is in no danger of becoming argument, let alone assertion, by claimed historical analogy. But it is notice of historical parallelism. A problem inherent in the very nature of empire, even of loose imperium such as that led by the contemporary United States, is that it proves impossible to know when to decide that the level of security currently attained can be judged sufficient. To go back to basics, one has to ask "Why might a threat be so designated?" Admittedly, there are some threats that lend themselves to metric analysis, or at least appear to do so. However, even in a case where the strategic mathematics could be grim, it cannot be denied that the answer to the most characteristic of strategist's questions, "So what?" may not be beyond all reasonable doubt. The point of note is that the real definition, especially with respect to the future, almost always is a zone for political deliberation and final judgment. In other words, definition of future threat can safely be assumed a matter allowing some discretion. Few threats are really self-defining. They do not emerge fully mature as to most likely consequences should they fail to be answered effectively today. The course of history, including history of a strategic kind, is a rat's nest of potential causes and possible effects. Definition of future threat has to include effort to discern possible and probable second- and third-order effects. Given that often it is distinctly challenging even to anticipate with tolerable accuracy what a first-order effect will be, it should be plain enough to see that threat definition is very much an art and, moreover, is one for which the military intelligence officer needs to be well prepared by rich strategic historical exposure.

Historical Parallels Are the Norm, Not the Exception.

Our contemporary lives oblige us to be alert to change, as great and small challenges that are new to us personally require attention. It is all a matter of perspective. It can be difficult even to appreciate the probable fact of historical parallels that should be of educational merit for us. The issues of the day press upon us, each seemingly sufficiently novel as to stretch beyond feasibility of significant alleviation by the application of methods long understood, well tried, and by and large true enough. An important theme for this analysis has been the eternal relationship between continuity and change. The reason American defense professionals should read and study Thucydides' admittedly challenging text is because his political and strategic world was not fundamentally different from ours.[44] This means that his book should be regarded as a work of rare insight on a subject contemporary to the author that, in its essential features, has not really altered for nearly 2 1/2 thousand years.

Of course, the differences between then and now are important, but they were changes that did not alter the nature of political relations among states by revolutionary impact or by cumulative adjustment over time. Indeed, it can be close to shocking to moral aspirations for our world, to be brought by facts to realize just how parallel the course of history often has been over millennia. The scope and depth of the historical parallelism typically is obscured from most of our notice by the necessary focus we have to place upon the challenges of today or upcoming tomorrow. Now inevitably has some novel features, novel in detail to us that is, though rarely so in prospective function. Because we have to be more or less genuinely expert in understanding our situation and its problems today, it is scarcely surprising that the true implications of the reality behind the concept of the great stream of time rarely achieve due recognition. Also, it is difficult to argue for the contemporary relevance of a text from ancient Greece, when the argument is advanced and sustained only by a small handful of scholars, whereas a wholly contemporary approach to our problem regularly is attended by a cast of hundreds or even thousands.

Perhaps the greatest challenge to clarity of understanding lies in the fact that Thucydides offers education rather than training. Much as Clausewitz wrote a book designed to help prepare soldiers to be competent in making the practical decisions they would have to make in war, so Thucydides wrote a generally pitiless history of the realities of disaster and triumph, genius and folly, and everything in between. The grand continuity that we find in the strategic dimension to the great stream of time is attributed largely to the persistence of just three elements over millennia: (1) a human nature with motives that do not change; (2) our exclusivity of political organization(s) to provide security; and (3) the logic of strategy that is an enduring reality, even though its practice frequently is not conducted competently.

American Exceptionalism Is a Half-Truth.

The exceptionalist literature, both scholarly and popular, is right in dwelling upon the ways in which the United States differs from all other polities on Earth—up to a point, at least. Unfortunately, no matter how elevated the moral or cultural and political the exceptionalist claims are believed to be (in America), the unforgiving nature of international politics applies no less to the United States than it does to other, lesser, polities. The United States is truly exceptional in its size, in the sheer weight of domestic assets that it brings to the game of nations, and as a direct consequence in the power as influence that it can exercise. The definition of future threat that is our organizing challenge to analyze and explain here probably should be understood not to derive significantly from those domestic features of the American experiment, of which we Americans understandably tend to be most proud. It is plausible to argue that our public values (and often private personal ones, also) add spice to the flavor of our foreign policy. But it is rather more plausible to argue that our great power, understood as relative influence, will continue to be the principal reason for our particular role in world political affairs.

The United States was the weightier side of an essentially bipolar balance all through the protracted Cold War, and we have remained preeminent, if now imminently challenged by China, for more than 2 decades thereafter. However, preeminence, even a relatively benign American form of hegemony, comes at a societal price that can be heavy. At the time of this writing, there is unmistakably convincing evidence that China is pursuing a generally prudent, cumulatively competitive course vis-à-vis the extant American-led order in East Asia and Asia–Pacific. The material and other evidence of competitive motivation is undeniable, and it will certainly continue to grow, provided domestic upheaval in China does not divert it. This Chinese competitive urge and long-term surge is all but obliged to be anti-American, simply because the United States remains the power still currently in the ascendant.

On similar logic, the new Russia can hardly avoid regarding the United States as an adversary, since the (American) security promise (or guarantee?) in Article V of the North Atlantic Treaty Organization Treaty of 1949 is the most essential element in the whole security pattern for (non-Russian) Eastern Europe. Russia is certain to provide ample evidence of threat in the future, but its relatively recent geopolitical losses and still worsening demographic situation in combination ought to damper risk-taking impulses in Moscow.

However, this author finds little or no comfort in such a seemingly reasonable judgment. Current Russian geopolitical, geostrategic, and demographic weaknesses ought to have an effect on policy that discourages adventure, but as recent and current events have demonstrated, this is not so.

Threat need not always flow strictly from adversaries ready enough to take crisis to the brink and beyond if the dynamics of confrontation assume an all but organic activism that escapes careful control. Definition of future threat has to take account of the possibility of the need to face down foreign leaders who are erratically and episodically apparently risk-tolerant to a dangerous degree. As a general rule, I must register a negative view of American efforts, no matter how well intended, to remake distinctly alien foreign societies with their richly un-American cultures. Physical menace sourced in such cultures certainly should be defined as future threat, but measures to control it should not entail forceful and protracted American endeavor to spread our ideas and practices.

My reason for this negative attitude is strictly pragmatic and is abundantly evidenced empirically. To be unmistakably explicit: the "American way" does not work well in many foreign cultures. If we have learned nothing else from the grim experiences in the 2000s in Afghanistan and Iraq, surely it must be this—it can never be sound strategy to attempt to achieve the literally impossible. The American imbroglios in Afghanistan and Iraq have been all too characteristic of mature hegemonic behavior, even misbehavior. Rephrased for the purpose of this monograph, it has been characteristic of very great powers, literally over the course of millennia, not to be sufficiently skilled in their statecraft and strategy as to know when and where to stop in the quest for an ever unreachable measure of national security. Threat definition is especially difficult at the present time because the United States plainly is well past the peak of its public's willingness to bear new strategic burdens, all the while a near-peer competitor in the form of the People's Republic of China assuredly is not going to be easily discouraged from opposing the existing security arrangements in Asia–Pacific. "Uneasy lies the head that wears the crown," is a summative comment appropriate to the subject here.

Chance, Black Swans, and Unknown Unknowns.

As something akin to a codicil to this whole discussion, it is advisable to remind readers of the limits to prudent rationality in threat definition and subsequent consequential defense preparation. I must warn that we should not be wholly unprepared for developments that may emerge as if by chance or by a process that cannot be anticipated or predicted. The point most in need of clearest registration is that we have to be ready enough to cope with surprising events that could have effects very damaging to our security for decades to come. An improbable, but certainly possible, practical alignment of Chinese and Russian policy goals and strategy could pose a gigantic continentalist challenge that we would find difficult to meet. In addition, we must worry about the threat that would be posed to our conception of international order by the world's first, but unlikely to be last, bilateral nuclear war. There will be many cases in the future, as there were in the past and are in the present (e.g., consider Gaza), of crises and wars in which the United States is not a principal participant; nonetheless, consideration of future threat definition cannot be indifferent to the incendiary potential of conflict that is almost structural to the basic dynamic functioning of the existing world order.

It is well to remember that there has been and remains a legacy of U.S. leadership following World War II. This is American in design, sustenance, and still substantially American in practice, even though the U.S. role has been increasingly muted in recent years. Prominent among the themes in Thucydides is the danger that can lurk underappreciated in the obligations that a hegemonic power finds it has to its security clients. This is a perennial cost that is an inalienable consequence of international political leadership.[45] The 21st may not be strictly another American century, but it is true to say that the most likely shape and structure of the international order yet to come is far from self-evident at present. This very uncertain international context makes a fundamental case for a prudent approach to threat definition and defense planning.

ENDNOTES

1. Klaus Knorr, "Threat Perception," in Klaus Knorr, ed., *Historical Dimensions of National Security Problems,* Lawrence, KS: University Press of Kansas, 1976, p. 116. This is the most insightful treatment of the concept and often arguable historical reality of threat known to this author.

2. Bernard Brodie, "The Continuing Relevance of *On War*," Carl von Clausewitz, *On War,* Michael Howard and Peter Paret, eds. and trans., Princeton, NJ: Princeton University Press, 1976, p. 54.

3. I have endeavored to explore and explain the mystery of the future, and especially the more characteristic errors in "fu-

turology," in my *Strategy and Defence Planning: Meeting the Challenge of Uncertainty*, Oxford, UK: Oxford University Press, 2014, esp. Chap. 1.

4. See Richard E. Neustadt and Ernest R. May, *Thinking in Time: The Uses of History for Decision Makers*, New York: Free Press, 1986.

5. I explored the problem of prudent defense planning in my *Defense Planning for National Security: Navigation Aids for the Mystery Tour*, Carlisle, PA: Strategic Studies Institute, U.S. Army War College, March 2014.

6. Thucydides, *The Landmark Thucydides: A Comprehensive Guide to the Peloponnesian War*, Robert B. Strassler, ed. and rev. trans., Richard Crawley (1876), New York: Free Press, 1996. The other widely available translation is that by Rex Warner, published by Penguin Books in 1954. It is a very good readable translation, but I must confess to a preference for the precision in Crawley's high Victorian English (somewhat amended) over the looser modern usage in Warner.

7. Williamson Murray, "Thucydides: Theorist of War," *Naval War College Review*, Vol. 66, No. 4, Autumn 2013, pp. 31-46.

8. Robert Gilpin, *War and Change in World Politics*, Cambridge, UK: Cambridge University Press, 1981, p. 7.

9. *Ibid.*, p. 227.

10. I have sought to initiate debate on anachronism among students of strategy, but potential debaters have yet to come forward. See Colin S. Gray, *The Strategy Bridge: Theory for Practice*, Oxford, UK: Oxford University Press, 2010, Appendix C, "Conceptual 'Hueys' at Thermopylae? The Challenge of Strategic Anachronism," pp. 267-277.

11. My thesis of there being an essential unity to all of strategic history is well supported in the outstanding essays in Williamson Murray and Richard Hart Sinnreich, eds., *The Past as Prologue: The Importance of History to the Military Profession*, Cambridge, MA: Cambridge University Press, 2006. Other books from the same

intellectual stable also merit notice: Williamson Murray, *War, Strategy, and Military Effectiveness*, Cambridge, MA: Cambridge University Press, 2011; and Williamson Murray and Richard Hart Sinnreich, eds., *Successful Strategy: Triumphing in War and Peace from Antiquity to the Present*, Cambridge, MA: Cambridge University Press, 2014. Also well worthy of an honorable mention is John France, *Perilous Glory: The Rise of Western Military Power*, New Haven, CT: Yale University Press, 2011, a major work marred seriously only by the author's near visceral disdain for some cultural explanation of strategic history.

12. I explain this process fully in my book, *The Future of Strategy*, Cambridge, UK: Polity Press, forthcoming in 2015.

13. See Harold D. Lasswell, *Politics: Who Gets What, When, How?* New York: Whittlesey House, 1936.

14. See Harry R. Yarger, *Strategy and the National Security Professional: Strategic Thinking and Strategy Formulation in the 21st Century*, Westport, CT: Praeger Security International, 2008.

15. By way of a rather sad example, General Stanley McChrystal's exemplary military career propelled him into the higher elevations where policy and strategy have to connect in mutually supportive fashion, but his many successful efforts in Iraq and Afghanistan were largely undone in subsequent practice by the inability of the contemporary American strategy making system to develop a strategy that could meet its political challenges. Readers of McChrystal's excellent memoirs can hardly help noticing that the author's superb tactical and operational skill-sets do not appear to have been matched at all closely by like competence in strategic appreciation. See Stanley McChrystal, *My Share of the Task: A Memoir*, New York: Penguin, 2013. McChrystal plainly was aware, even acutely so, of the difficulty in translating tactical success into lasting strategic advantage. For example, he commented as follows on his special operations: "We executed missions; we did not wage campaign" (p. 181). A little further on, he observes that "[w]ithin a couple of months, by October [2005], we were able to see real evidence of the strategic impact I hoped would materialize when I explained it to the operators in Al Qaim that day" (p. 186). But, without an inclusive and single understanding of the political purpose of the campaign, it was not pos-

sible to devise a strategy that might enable the desired political path to be followed. Hew Strachan, "Strategy or Alibi? Obama, McChrystal and the Operational Level of War," *Survival*, Vol. 52, No. 5, October-November 2010, pp. 157-182, is very much to the point. Generals will wage war tactically, they may succeed in performing operationally, but they cannot do strategy in the absence of political direction worthy of the name.

16. When actual combat performance reveals a large adverse tactical gap between the fighting power of our soldiers and that demonstrated by those of the enemy, operational artistry or strategic skill will not stage a rescue for you from acute and even dangerous embarrassment. As a recent biography of Dwight Eisenhower explained: "The fact is the green American troops were no match for the German veterans" [in Tunisia in February 1943]. Jean Edward Smith, *Eisenhower in War and Peace*, New York: Random House, 2012, p. 262. This judgment is supported amply in Rick Atkinson, *An Army at Dawn: The War in North Africa, 1912-1943*, New York: Henry Holt, 2002, pp. 344-347.

17. Raymond Aron, *Peace and War: A Theory of International Relations*, New York: Doubleday, 1966, p. 285.

18. The concept and alleged empirical reality of America's "unipolar moment" was given a hard time both by liberal scholars and by realists who were suitably fearful of the danger of an Athenian moment when "victory disease" all too easily could override prudential reasoning. For some appropriate views, see Charles Krauthammer, "The Unipolar Moment Revisited," *The National Interest*, Vol. 70, 2002-03, pp. 5-17; and Christopher Layne, *The Peace of Illusions: American Grand Strategy from 1940 to the Present*, Ithaca, NY: Cornell University Press, 2006. For a brief overview of the "unipolar" 1990s, see Colin S. Gray, *War, Peace and International Relations: An Introduction to Strategic History*, 2nd Ed., Abingdon, UK: Routledge, 2012, pp. 248-250.

19. Thucydides leaves us no doubt that the principal threat to the security of Athenians flowed more from the distinctly flawed working of the empire's democratic politics, especially its proclivity to promote crowd pleasing demagogues who were short of competence, high ethical standards, or both, than from vengeful Persians or strategically pedestrian Spartans. Political ruin tends

to begin and end at home. Students of international relations need to remember this plain warning from the historical record.

20. I choose to define grand strategy as "the direction and use made of all the assets of a security community, including its military instrument, for the purposes of policy as decided by polities." Gray, *The Strategy Bridge*, p. 263. This definition is, of course, perilously inclusive, possible even to a fault. An outstanding brief discussion of the concept of grand strategy, particularly in its relationship to history, has been provided by scholarly historian John Gooch. He advises that:

> [b]ecause grand strategy is the pursuit of national political objectives by harnessing all the means at the disposal of the state, politics lies at its core. This fact alone accounts for the longevity of Clausewitz, whose timeless observations still command the respect and admiration of most students of strategy, and for the obvious fervor with which his adherents stress the political side of strategy over the technical and technological.

John Gooch, "History and the Nature of Strategy," in Murray and Sinnreich, eds., *The Past as Prologue*, p. 143. There has been some rise in American interest in the concept of grand strategy, but this very large idea continues to defy attempts to secure a consensus as to its most proper meaning. Williamson Murray, Richard Hart Sinnreich, and James Lacey, eds., *The Shaping of Grand Strategy: Policy, Diplomacy and War*, Cambridge, MA: Cambridge University Press, 2011; and Charles Hill, *Grand Strategies: Literature, Statecraft, and World Order*, New Haven, CT: Yale University Press, 2010, are noteworthy attempts to provide historical and cultural contexts for the concept, while the most scholarly investigation to date is Lukas Milevski, "The Modern Evolution Grand Strategic Thought," Ph.D. diss., Reading, UK: University of Reading, 2014, though Milevski, on balance, may be deemed more to have compounded controversy over meaning, rather than help solve it. However, his careful study is a most important advance toward our ability to find proper use of this vital idea.

21. This episode is too recent for there to be much impressive theoretical literature about it. My own distinctly humble efforts at explanation of the meaning of contemporary technical develop-

ments were summarized in a work and a monograph: *Strategy for Chaos: Revolutions in Military Affairs and the Evidence of History*, London, UK: Frank Cass, 2002; and *Making Strategic Sense of Cyber: Why the Sky is Not Falling*, Carlisle, PA: Strategic Studies Institute, U.S. Army War College, April 2013. Williamson Murray, "Thinking About Revolutions in Military Affairs," *Joint Force Quarterly*, Vol. 16, Summer 1997, pp. 69-76, is a rare outstanding treatment of the revolution in military affairs (RMA) enthusiasm of the period. The problem with the digital revolution is that it has occurred, indeed is still occurring, during a quarter century blessedly bereft of great power war, meaning that technical and tactical advances have galloped on, leaving strategic understanding largely shrouded in mystery.

22. The proposition that politics/policy and strategy are more important than tactics and technology has become close to thematic in the writings of historian Williamson Murray. See his *Military Adaptation in War: With Fear of Change*, Cambridge, MA: Cambridge University Press, 2011; Murray, *War and Military Effectiveness*, Cambridge, MA: Cambridge University Press, 2011; and Murray and Sinnreich, eds., *Successful Strategies*, Cambridge, MA: Cambridge University Press, 2014.

23. Intellectual mastery of the subject of nuclear threat requires understanding of the somewhat rival theories and arguments advanced by both Thomas C. Schelling and Herman Kahn. See Thomas Schelling, *The Strategy of Conflict*, New York: Oxford University Press, 1963; Thomas Schelling, *Arms and Influence*, New Haven, CT: Yale University Press, 1966; Herman Kahn, *On Thermonuclear War*, Princeton, NJ: Princeton University Press, 1960; and Herman Kahn, *Thinking About the Unthinkable*, New York: Avon Books, 1962. The sharp contrast between Schelling and Kahn's thinking on nuclear threat is an intellectual centerpiece to Keith B. Payne, *The Great American Gamble: Deterrence Theory and Practice From the Cold War to the Twenty-First Century*, Fairfax, VA: National Institute Press, 2008.

24. A careful set of historical case studies concluded that reputation for integrity toward commitments has not been the dominant force for credibility. A superior weight of evidence supports the proposition that political leaders are far more impressed by their assessment of the quality of current interest an opponent ap-

pears to have in a particular issue, rather than in that adversary's past record of fidelity in promised action to support declared interests. See Daryl G. Press, *Calculating Credibility: How Leaders Assess Military Threats*, Ithaca, NY: Cornell University Press, 2005. This is one of those modest insights from scholarship that can have immense practical significance. It is no exaggeration to claim that the whole theory of deterrence in strategy may need to be redrafted if Press is judged to have read history correctly. It should be needless to say that it is orthodox to believe that reputation tomorrow can be at risk to perception of irresolution and weakness today.

25. Although states and their societies have armed forces that reflect more than casually their social roots and sustaining influences, also it is true that war, indeed each particular war, is a whole phenomenon that more than somewhat evolves seemingly organically, changing the style and some of the practices of the competitive belligerents. My point is that despite the hostility involved, a war is a dynamic episode that adversaries share. To a meaningful degree, they must wage a single contest, and their means and methods are certain to share some common features. For a particularly persuasive explanation of this character of war, see Patrick Porter, *Military Orientalism: Eastern War Through Western Eyes*, London, UK: Hurst, 2009, pp. 65, 170, 191.

26. At long last, the ideas of adaptability and adaptation have begun to be recognized for the importance that they must have. See Murray, *Military Adaptation in War.*

27. See Colin S. Gray, "Strategic Thoughts for Defence Planners," *Survival*, Vol. 52, No. 3, June–July 2010, pp. 36–38.

28. An outstanding and revealing recent report on command experience in American context is General Stanley McChrystal, *My Share of the Task*, New York: Penguin, January 2013.

29. Richard Ned Lebow and Barry S. Strauss, eds., *Hegemonic Rivalry: From Thucydides to the Nuclear Age*, Boulder, CO: Westview Press, 1991, aired some rival views of Thucydides' relevance to our modern times, while a rather later assessment by David A. Welch at least was admirably explicit in to its message to scholars: "Why International Relations Theorists Should Stop Reading

Thucydides," *Review of International Studies*, Vol. 29, No. 3, July 2003, pp. 301-319.

30. Thucydides, *The Landmark Thucydides*, p. 16.

31. The most helpful guide to the subject of this monograph is Michael Howard, *The Lessons of History*, New Haven, CT: Yale University Press, Chap. 1.

32. As Zara Steiner makes very clear indeed, the United States was quite heavily engaged in international relations during the 1920s and 1930s, but by choice it was not really on the field of play that contained the elements we must include as being relevant in the category of strategic history. See the magisterial treatment of the interwar decades in her books: *The Lights That Failed: European International History, 1919-1933*, Oxford, UK: Oxford University Press, 2005; and *The Triumph of the Dark: European International History, 1933-1939*, Oxford, UK: Oxford University Press, 2011.

33. Knorr, "Threat Perception," p. 112.

34. See Nassim Nicholas Taleb, *The Black Swan: The Impact of the Highly Improbable*, New York: Random House, 2007.

35. McChrystal provides an excellent participant's view of recent history of the U.S. Army.

36. Thucydides' core arguments on strategy are well summarized in Athanassios G. Platios and Constantinos Koliopoulos, *Thucydides on Strategy: Grand Strategies in the Peloponnesian War and their Relevance Today*, London, UK: Hurst, 2010. An essay helpfully contextual for the book by Platios and Koliopoulos is Paul A. Rahe, "Thucydides as Educator," in Murray and Sinnreich, eds., *The Past as Prologue*, pp. 95-110.

37. Michael Howard advises that:

> history, whatever its value in educating the judgement, teaches no 'lessons,' and the professional historians will be as skeptical of those who claim that it does as professional doctors are of their colleagues who peddle patent medicines guaranteeing instant cures. The past is infinitely various,

> an inexhaustible storehouse of events from which we can prove anything or its contrary. . . . historians may claim to teach lessons, and often they teach very wisely. But 'history' as such does not.

Lessons of History, p. 11.

38. I advance this claim on behalf of historical education in my book, *Strategy and Defence Planning.*

39. The concept of the historical parallel avoids the impossible difficulties with analogy, while being encouraging of potentially fruitful search for enlightening similarities in the structure of diverse historical episodes. I am pleased to acknowledge debts to Klaus Knorr and Michael Howard for my appreciation of the merit in this idea. Knorr, "Threat Perception"; Howard, *The Lessons of History*, Chap. 1.

40. See the following small gem of a book, General Sir Andrew Skeen, *Lessons in Imperial Rule: Instructions for British Infantrymen on the Indian Frontier*, (1932) London, UK: Frontline Books, 2008.

41. Donald Rumsfeld, *Known and Unknown: A Memoir*, New York: Sentinel, 2011, p. xiii.

42. Thucydides, *A Comprehensive Guide to the Peloponnesian War*, p. 43.

43. *Ibid.*, pp. 56, 128.

44. No less an authority than General of the Army George Catlin Marshall told an audience at Princeton University on February 22, 1947, that he doubted:

> whether a man can think with full wisdom and with deep convictions regarding certain of the basic international issues today who has not at least reviewed in his mind the period of the Peloponnesian War and the Fall of Athens.

Quoted in Paul A. Rahe, "Thucydides as Educator," in Williamson Murray and Richard Hart Sinnreich, eds., *The Past as Prologue*, p. 99.

45. See Thucydides, *A Comprehensive Guide to the Peloponnesian War*, pp. 53–54. Because it rests largely on an historical narrative structure, Thucydides' great book is not designed in the manner of our contemporary social science text books, with chapters discretely targeted to develop particular ideas or theories. As a consequence, one finds a particular phenomenon (e.g., alliance behavior and misbehavior) integrated into the whole of the narrative, which is to say in proper historical context. Consideration of alliance or client, relationships in Thucydides is assisted usefully by the following works: Donald Kagan, *Thucydides: The Reinvention of History*, New York: Viking, 2009; and Platios and Koliopoulos, *Thucydides on Strategy*. While Athens was a strategically giant ally, it was inevitably a greatly resented collector of security tribute, too. On balance, both then and now, it is a persisting reality of international politics that the dominant partner in a relationship of security dependency finds that it must adopt some local causes of concern, ambition, and anxiety in the interest of tolerable alliance relations. Small security-dependent clients are fully capable of manipulating a very great power into an imprudent commitment to provide assistance.

U.S. ARMY WAR COLLEGE

Major General William E. Rapp
Commandant

*****

STRATEGIC STUDIES INSTITUTE
and
U.S. ARMY WAR COLLEGE PRESS

Director
Professor Douglas C. Lovelace, Jr.

Director of Research
Dr. Steven K. Metz

Author
Dr. Colin S. Gray

Editor for Production
Dr. James G. Pierce

Publications Assistant
Ms. Rita A. Rummel

*****

Composition
Mrs. Jennifer E. Nevil

Made in the USA
Monee, IL
06 October 2024

67251252R10042